Occasional Paper No. 43

Variability in tender bids for refurbishment work

Lee Kiang Quah, PhD, MCIOB, MBIM

The Chartered Institute of Building
Englemere, Kings Ride, Ascot, Berkshire SL5 8BJ

PREFACE

This publication presents the main findings of an investigation into the reasons for the higher variability in the tender bids for refurbishment work as compared to new build. The main objective of the study was to determine the main areas of variability in the tenders submitted by competitors. The study, carried out at the Department of Building, Heriot Watt University, was funded in part by a QEII Scholarship awarded by The Chartered Institute of Building.

ACKNOWLEDGEMENTS

I am grateful to the Builders' Conference UK for providing an Educational Trust Fund Award towards this study and for donating the primary data base. Special thanks are also due to the directors and chief estimators of the 40 contractors who despite their heavy schdules gave generously their time, expertise and donated the secondary data base.

1. VARIABILITY IN TENDER BIDS FOR REFURBISHMENT WORK

DESCRIPTION OF THE TENDERS INVESTIGATED

A main sample consisting of 1,350 projects spanning the period January 1984 to August 1988 was obtained from the tender reporting service of the Builders Conference.

The project sizes ranged from less than £10,000 to over £3 million. 60% of the 1,350 projects were below £500,000/ in value.

The following variables which could affect tender value, and hence tender bid variability, were selected for investigation:

(a) date of tender;
(b) client type;
(c) project size;
(d) project type;
(e) project location;
(f) size of bidding set.

MEASURES OF DISPERSION IN TENDER BIDS

For simplicity, the measure of dispersion in the tender bids used was the range, from:

Range = $\frac{\text{(Highest Bid} - \text{Lowest Bid)}}{\text{Lowest Bid}}$ × 100% for each project.

The average of the ranges obtained from the 1,350 projects was 19.9%.

VARIABLES WHICH SIGNIFICANTLY AFFECT THE DISPERSION OF TENDER BIDS

Tenderers can expect that as the size of a project increases and/or the number of bidders increase in a project, the dispersion in the tender bids would ~~increase~~ DECREASE. This is due to the effect of increased competition; in particular competition between more evenly matched sized firms in the larger projects.

No other variables were found to significantly affect the bidding dispersion.

WIN MARGIN

The win margin was computed from:

Win margin = $\frac{\text{(Second Bid} - \text{Lowest Bid)}}{\text{Lowest Bid}}$ x 100% for each project

The average of the win margins for the 1,350 projects represents the 'money left on the table' for refurbishment projects.

The average ~~bid spread~~ WIN MARGIN of refurbishment tenders was 5.7%.

50% of the 1,350 projects had win margins of 3.8% and over. There is thus an even chance that contractors tendering for refurbishment work can increase their tenders by 3.8% without necessarily affecting their chance of being the lowest bidder.

VARIABLES WHICH SIGNIFICANTLY AFFECT THE WIN MARGIN OF THE TENDER

Tenderers should expect that as the size of the project increases and/or as the number of bidders increase in a project, the win margin should decrease. This is due to the effect of competition as explained earlier.

No other variables were found to affect significantly the win margin.

TRUE DISPERSION AND WIN MARGIN OF TENDERS

A secondary sample of 75 projects, drawn from the main sample of 1,350 projects, was investigated to determine the 'true' dispersion and win margins in the tenders. The 'true' quantitative measures are obtained from the dispersion in the actual quantum of priceable work within the tenders, ie. the tender sum after deducting all tenderers common fixed costs of prime cost, provisional and contingency sums. As such, the 'true' measures portray the bidding performance of tenderers more accurately.

$$\text{True range of tenders} = \frac{\text{Highest Bid} - \text{Lowest Bid}}{\text{Lowest Bid} - (\text{PC}+\text{Prov.}+\text{Contin}) \text{ Sums}} \times 100\% \text{ for each project}$$

$$\text{True win margin} = \frac{\text{Second Bid} - \text{Lowest Bid}}{\text{Lowest Bid} - (\text{PC}+\text{Prov.}+\text{Contin}) \text{ Sums}} \times 100\% \text{ for each project}$$

The average of the true ranges for the 75 projects was found to be 30.3% and that for the true win margins 9.0%.

VARIABLES WHICH SIGNIFICANTLY AFFECT THE TRUE DISPERSION AND WIN MARGINS OF THE TENDERS

Tenderers can expect that as the proportion of prime cost, provisional and contingency sums in a tender increase, the true dispersion on the tender bids, as well as the true win margin, will increase. This is because the quantum of priceable work reduces. The true dispersion and the true win margin will serve more as a measure of the relative efficiencies of competing firms, this efficiency being reflected in the mark-up and preliminaries cost, which will then form the main bulk of priceable work.

The effect of increased dispersion and increased 'win margin sums' in the tender bids with increased PC sums, provisional and contingency sums will hence be particularly emphasised in projects where uneven sized firms meet in competition.

2. COMPETITIVE PATTERNS IN REFURBISHMENT WORK

DESCRIPTION OF THE FIRMS INVESTIGATED

40 collaborating firms were studied as detailed in Table 1.

Table 1

Size of firm	Turnover in construction activities	No. in study
Small	$\leq$ £20M	13
Medium	$>$ £20M $<$ £75M	17
Large	$>$ £75M	10

Type of firm	Turnover in refurbishment work	No. in study
General	$<$ 50% in refurbishment work	31
Specialist	$>$ 50% in refurbishment work	9

Three measures were used to compare the bidding performance of the contractors viz:

(a) Success rate

This was obtained by dividing the number of tenders won by the number of tenders submitted and expressing this ratio as a percentage. (The lowest tender is assumed to be the winning tender). This measure could be affected by the number of bids which are, for a number of reasons, not submitted with any anticipation of success. This success rate is compared with the 'success rate by chance' to determine whether the contractor has beaten the odds of winning a tender. The 'success rate by chance' is computed by dividing the number of tenders submitted by a contractor by the number of tenderers.

(b) Win margin

This is the margin by which the contractor wins a project. To enable it to be compared between projects, it was standardised by the lowest bid, ie. the winning bid of the contractor concerned.

(c) Failure margin

This measures the difference between a contractor's unsuccessful bid and the lowest bid in that project, ie. the margin by which a job was lost. To enable the failure margin to be compared between projects, it was standardised by the lowest bid in each project.

COMPETITIVE PERFORMANCE OF DIFFERENT SIZED FIRMS

Table 2 shows the comparative competitive performance of contractors grouped by size of firm. The three performance measures were weighted with a factor which was the number of tenders with which each contractor was concerned.

Table 2 Comparative competitive performance of contractors grouped by size of firm

Size of firm (No. in group)	No of specialists in group (%)	Success rate		mean weighted win margin	mean weighted failure margin
		Mean weighted success rate	Differences between success rate and win by chance		
Small (13)	1 (7.7)	20.26	0.43	6.46	10.75
Medium (17)	4 (29.4)	18.70	-0.40	4.71	9.80
Large (10)	3 (30.0)	19.02	2.98	4.55	9.57

Although Table 2 shows that there are minor observable differences between the three groups of firms, tests indicate that these differences were not statistically significant.

To further amplify the comparative competitive performance of the three groups of contractors, a scatter plot of the individual performance of contractors in each group is shown in Figure 1.

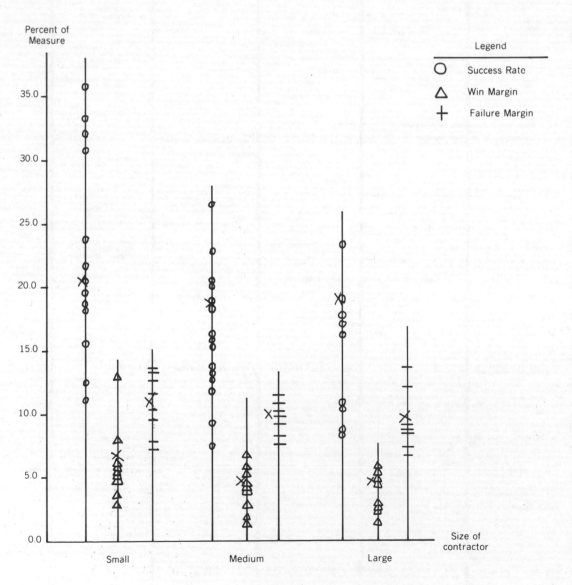

Figure 1　　Comparative Competitive Performance of contractors by size of firm

It can be seen from Figure 1 that although small firms had the highest mean weighted success rates, the relative performance of the 13 firms in the group was the most variable, as evidenced by scatter of their success. On the other hand, the ten firms within the large size category showed greater consistency in individual performances.

It should be noted that the competitive performance of the groups of firms could be influenced by the number of specialist in the group, eg. the large sized firms contained the greated proportion of specialist refurbishment contractors. This could contribute to the better competitive performance.

COMPETITIVE PERFORMANCE OF DIFFERENT FIRM TYPES

Table 3 Comparative competitive performance of contractors grouped by firm type

Type of firm (No. in group)	Success rate		Mean weighted win margin	Mean weighted failure margin
	Mean weighted success rate	Difference between success rate and win by chance		
Specialist contractor (9)	21.45	2.25	4.92	10.07
General contractor (31)	16.65	-0.84	5.05	9.84

It can be seen from Table 3 that specialist refurbishment contractors perform significantly better than general contractors in tendering. Figure 2 also shows that the variability in the performance measure of individual refurbishment contractors was lower than that of the general contractors.

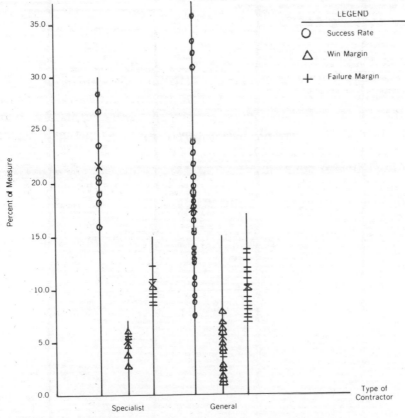

Figure 2   Comparative competitive performance of contractor's by type of firm

5

COMPETITIVE MIX IN TENDERING

The competitive mix in terms of types of firms and size of firms competiting was examined for a sample of 75 projects.

Competitive mix in terms of type of firms

For all the seventy-five projects, there was at least one specialist refurbishment contractor amongst the competitors within a project, two or three being more the norm. In only four (5.3%) of the seventy-five projects were all competitors specialist refurbishment contractors. There were no projects where all of the competitors were general contractors.

From the results of the analysis on the comparative bidding performance of refurbishment and general contractors it can be concluded that general contractors tendering for refurbishment work will, in all probability, be out bidded by refurbishment specialists within the group of competitors.

Competitive mix in terms of size of firms

In seventy-four (98.7%) of the seventy-five projects, the competitors came from different sized firms. In fifty-three (71.6%) of these seventy-four projects, the competitive mix was between competitors in the next incremental group size, ie. small and medium or medium and large. In fifteen (20.3%) of the seventy-four projects, the competitors came from the three sizes of firms. In the remaining six (8.1%) projects, the competitors came from small and large firms only.

It was also noted that there was more than one large contractor competing with small and medium sized contractors in each of the ten (83.3%) out of twelve projects in the size range of between £100,000 and £250,000. However, in the seven projects in the range above £1,500,000 there were only two (30%) projects where a small sized firm was in competition with large and medium sized firms. The two small firms were specialist refurbishment companies which had been acquired by large sized firms, and had gained access to the tender list because of their parent companies.

Although the results of the statistical analysis on the competitive performance of the different sized firms showed that their success rate was not significantly different, in practice this may not be so. The presence of large firms competing for projects, which are traditionally the domain of small firms, could affect the competitive behaviour of the tenderers. The impact of a competitive mix of unequal sized firms was stronger in projects where the prime cost, provisional and contingency sums were high, as it leaves the competitors to compete mainly on mark-up and preliminaries. In such situations, small firms may not be able to bid competitively against large firms, who will probably have a better competitive advantage.

3. DIFFERENCES IN METHODS OF ESTIMATING AND TENDERING AND RATIONALISATION FOR COMPARATIVE ANALYSIS OF TENDER BUILD-UPS

In examining the tender documentation supplied by the collaborating 40 firms for the purpose of determining the variability in the tender bid components of competing contractors, several differences in the methods of estimating and tendering were identified.

These differences had to be rationalised in order to enable a true comparison to be made of the actual variations in the cost of the components of the tender. The rationalisation process was carried out in favour of the more common methods of estimating and tendering. All data of contractors who differed from these methods were rationalised accordingly.

The following summarise the main differences in methods of estimating and tendering identified from the 229 tender bids which were analysed, and the method of rationalising the data.

ESTIMATING PRELIMINARIES/PROJECT OVERHEAD COST

The exact number of sub-components in the Preliminaries section varies substantially between contractors, ranging from a minimum of twelve to a maximum of thirty-one. (The CIOB's *Code of Estimating Practice* identifies a total of fifteen sub-components).

Variability in Preliminaries cost was compared under seven main heads: staffing; plant; scaffolding; temporary accommodation; temporary supplies and services; general labour and attendances; insurance and bonds. All other items were grouped and compared separately.

ATTENDANCES ON NOMINATED WORK AND GENERAL LABOUR

Thirty-eight out of the forty-two contractors did not price general and special attendance against the prime cost item in the bill. Instead, the cost of attendance on nominated sub-contractors was included within the Preliminaries under the heading of General Labour and Attendances. This was the cost for a calculated number of labourers needed for clearing, cleaning, unloading and general attendance on all sub-contractors, as well as the contractor's own work.

The remaining four contractors priced the cost of both general and special attendances under the respective items within the bill. The number of general labourers needed for the entire project was also estimated within the Preliminaries. The cost was then balanced against the aggregated cost of attendances allowed in the respective work items. Any shortfall between the sums was then added in the Preliminaries.

To rationalise these two differing methods of pricing attendances, the cost of this item was transferred from the Prime Cost section of the bill to Preliminaries in the case of the four contractors concerned.

PROFIT FROM NOMINATED WORK

Thirty-eight of the forty-two contractors did not price the profit item in the nominated work section of the bill. Instead, the profit was expressed as a mark-up applied to the total nett estimate (of which nominated work formed part) in the tender summary sheet. However, two of the thirty-eight contractors chose to price profit on nominated work separately from the mark-up applied to the rest of the nett estimate. The other four contractors priced profit within the relevant item in the bills.

In rationalising these different methods, all sums of money in respect of profit on nominated work, whether inserted within the bill or separately in the tender summary sheet, were transferred to the mark-up component of the bid.

ALLOCATION OF GENERAL OVERHEADS

The following differences were detected in the methods of allocating general overheads costs.

Site surveying

The cost of employing site quantity surveyors to service all projects undertaken by the company was borne as a general overhead cost by thirty-nine of the forty-two contractors. The remaining three contractors included this cost as a project overhead in the Preliminaries. However, two contractors from the former group adopted the latter method of cost allocation during the progress of the study.

To rationalise these differences, cost in respect of site surveying, which was priced within Preliminaries, was transferred to the mark-up component of the tender bid.

Insurances

Two of the forty-two contractors subscribed to a blanket insurance policy which covered all insurable liabilities in respect of their trading operations. The premium was based on their annual turnover and was allocated as a general overhead cost. The other contractors costed insurance premiums for each project and included this either in the Preliminaries or as a separate item in the tender summary.

In rationalising these three methods of pricing insurances, the two contractors who took the blanket policy were asked to give an estimate of the cost of insurances for the project, assuming the policy was taken out separately as with the other contractors. This sum was then deducted from the mark-up component of their tender bid and transferred to the Preliminaries under a sub-component entitled insurances and bond. For contractors who priced the cost in the tender summary, the sum involved was transferred to Preliminaries.

DISCOUNTS RECEIVABLE

Thirty-eight of the forty-two contractors netted their cost estimates by the value of the discounts receivable from nominated and domestic sub-contractors and suppliers. One contractor was in the habit of retaining half of the discounts, while the remaining three contractors retained the full discounts receivable.

It is believed that the quantum of discounts represents profit, which the contractor is prepared to forego if he decides to nett his cost estimated by the same sum. This view was supported by thirty-four (72.9%) of the collaborators; the other twelve (26.0%) were of the view that discounts were related to nett costs and not profit.

Adopting this view and the views of the majority of the collaborators, the rationalisation of these differing methods of dealing with discounts was undertaken in the following manner.

A new component termed 'anticipated return' was introduced for comparing the bids. This component is not an established priceable component: it represents the quantum of general overheads and profit contractors anticipate receiving in excess of the mark-up indicated in the tender. Comparisons of the nett estimate were based upon the gross figures as priced within the tender bills. Contractors who gave up discounts by netting their estimates had the discounts deducted from their mark-up component, thereby reducing their profit by the same amount. Their 'anticipated return' was thus the same value as their adjusted mark-up. For contractors who retained half or full discounts, the mark-up shown in their tender bids were not adjusted. Their 'anticipated return' would then be their unadjusted mark-up figure, plus the value of the discounts retained.

RISK ADDITIONS AND SCOPE DEDUCTIONS

Only two of the forty-two contractors had a specific heading of risk/scope within their tender summary sheets. A sum of money was either added or deducted from the tender (ie. from the sum derived after the mark-up had been applied to the nett estimate), producing a revised sum which was then offered as the tender price.

On a small number of occasions, two contractors inserted risk allowances in their Preliminaries costs under the heading of 'other costs'.

One contractor had a very detailed tender summary sheet showing computations for scope deductions in respect of measurement gains and building price index gains.

Two other contractors had an item entitled 'Add or Deduct at Director's Discretion' within their tender summary sheets. The reasons for the addition or subtraction sum was not always indicated, although common comments included risk or scope in Provisional Items and Sums.

The balance of the thirty-five contractor's tender summary sheets made no specific reference to risk or scope. They presumably accommodated risk/scope within the estimating and tendering process through the mark-up on their nett estimate.

As a method of rationalisation, all risk allowances costed by the contractors concerned were transferred from their respective positions within the tender to the mark-up component, indicating that a higher mark-up would be required for the risk to be carried.

Scope deductions in the tender were adjusted by deducting this sum from the mark-up shown in the tender summary. It was then, in turn, reflected in the anticipated return, showing that the contractor actually anticipated the value of this scope reduction to the reimbursed by way of claims or other means.

MARK-UP APPLIED

Forty of the forty-two contractor's mark-up addition included an allowance for general overheads and profit. It was not possible to determine the breakdown for these two sub-components of the mark-up. Mark-up was generally applied to the total nett estimate after deducting discounts, daywork, provisional and contingency sums, the latter two items being deemed to contain their own overhead and profit margins.

However, one contractor, showed separate and differing mark-ups on nominated and domestic sub-contractor's work, as well as on his own work. His separate mark-ups were aggregated together as a single mark-up figure to align with the others.

Another contractor applied a substantially higher general overhead allowance to the estimated labour cost component of his own work, and varying profit margins to the remainder, including preliminaries and domestic and nominated sub-contractor's work. In this case, the overhead allowance and the separate profits were aggregated into a single mark-up figure for comparison with the rest of the forty contractors.

4. PROFILE OF A LOWEST REFURBISHMENT TENDER

Figure 3 shows how the tender was broken down into components of cost for the purpose of analysis.

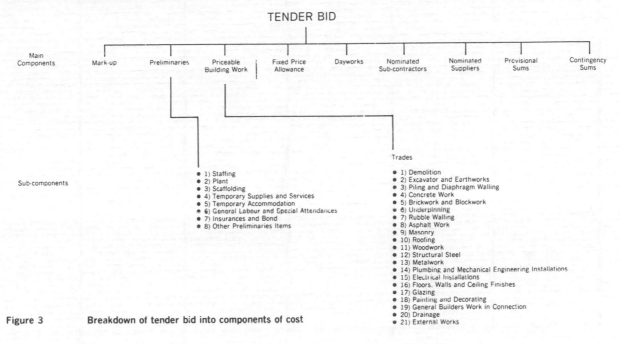

Figure 3   Breakdown of tender bid into components of cost

PROFILE OF THE MAIN COST COMPONENTS OF THE TENDER

The detailed tender build-ups of competitors for fifty-five projects were used to ascertain the proportion of cost of each component in a refurbishment winning bid. Table 4 shows the range of the proportion of cost of each component from which it can be seen that it varied widely across the fifty-five projects.

Table 4 Proportion of cost of main components in a refurbishment tender

Main components of tender	Proportion of cost of component in tender (%)		
	Maximum	Mean	Minimum
Mark-up	9.0	2.9	-3.5
Preliminaries	20.0	9.5	2.4
Builder's own work	35.8	12.1	0.6
Domestic sub-contractors	75.6	21.6	2.1
Fixed price allowance	2.3	0.6	0.5
Daywork	8.8	1.9	0.4
Nominated sub-contractors	76.6	39.0	4.5
Nominated suppliers	5.4	0.8	0.7
Provisional sums	39.4	8.8	0.6
Contingency sums	7.7	2.8	0.7
	Base	100.0	

The mean values of the proportion of the cost components within the tender were used to determine the cost profile of the nine main components of the tender. Figure 4 shows the profile of a winning tender bid in terms of the components. It can be seen that nominated sub-contractor's work occupied the highest proportion of cost in the tender at 39%. Refurbishment projects were won on a (mean) mark-up comprising 2.9% of the tender sum (equivalent to a mark-up of 3% applied to the nett estimate). The Anticipated Return, on the other hand, was 3.7% of the tender sum.

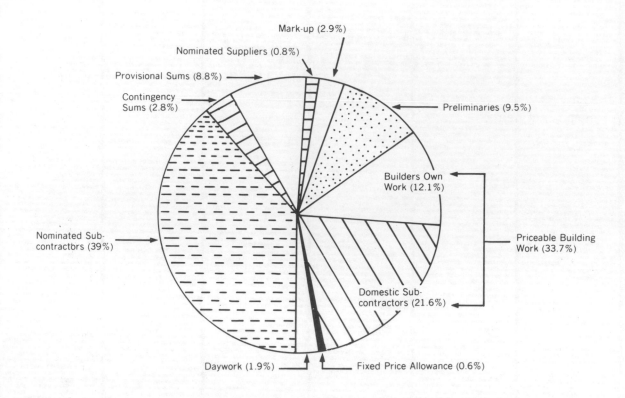

Figure 4   Profile of lowest bid for a refurbishment project

PROFILE OF THE COST COMPONENTS OF PRELIMINARIES IN A TENDER

The proportions of the cost of the sub-components within the main components of Preliminaries and the Priceable Building Work was also computed. Table 4 shows the range of the cost proportions of the eight sub-components of the preliminaries in the fifty-five projects. The variability in the proportions of the cost of sub-components within preliminaries is evident from the Table. Table 5 shows the range of the cost proportions of the eight sub-components of the preliminaries in the fifty-five projects.

Table 5 Proportion of cost of sub-components in preliminaries

Sub-component of preliminaries	Proportion of cost of sub-component in preliminaries (% of total preliminaries cost)		
	Maximum	Mean	Minimum
Staffing	44.8	31.2	14.3
Plant	28.0	10.0	1.6
Scaffolding	45.6	14.5	0.6
Temporary supplies & services	15.3	6.2	0.90
Temporary accommodation	17.0	3.9	0.30
General labour and special attendance	38.7	22.0	4.4
Insurance and bond	17.5	4.8	0.1
Others	40.1	7.4	0.1
	Base	100.0	

The mean value of the cost proportion of the sub-components (see Table 5) was used to determine the cost profile of the eight sub-components of the Preliminaries. This is shown in Figure 5. The highest proportion of Preliminaries cost was in the Staffing sub-component at 31.2%.

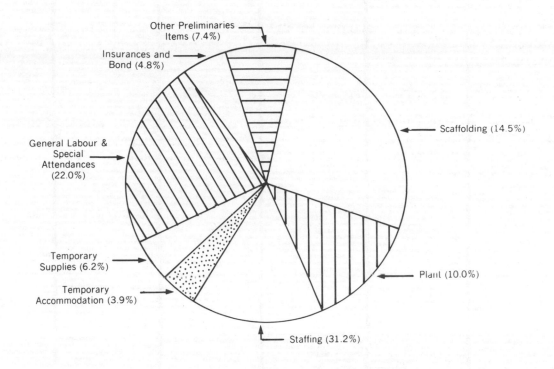

Figure 5   Cost profile of sub-components of preliminaries

PROFILE OF THE MAIN TRADES IN A TENDER

Table 6 shows the range of the cost proportion of the twenty-one trade sub-components of Priceable Building Work in the fifty-five projects.

Table 6 Proportion of cost of each trade sub-component in the priceable building work

Trade sub-component of Priceable Building Work	Proportion of cost of sub-component in the Priceable Building Trades Work (% of the total cost of Priceable Building Trade Work)		
	Maximum	Mean	Minimum
1) Demolition	58.8	17.3	1.5
2) Excavator & earthworks	10.5	1.1	0
3) Piling & diaphragm walling	12.4	0.3	0
4) Concrete work	28.6	4.7	0
5) Brickwork & blockwork	16.4	4.2	0
6) Underpinning	17.6	0.9	0
7) Rubble walling	13.4	0.2	0
8) Masonry	14.6	0.6	0
9) Asphalter work	7.5	0.6	0
10) Roofing	51.9	3.7	0
11) Woodwork	45.0	13.4	0
12) Structural steelwork	15.3	2.3	0
13) Metalwork	13.0	7.3	0
14) Plumbing & mech. eng installation	56.0	6.4	0
15) Electrical installations	21.7	1.1	0
16) Floor, walls & ceiling finishing	82.2	21.2	0
17) Glazing	55.3	1.9	0
18) Painting & decoration	35.0	7.1	0
19) General builder's work in connection	27.9	3.1	0
20) Drainage	11.7	1.1	0
21) External work	29.3	1.5	0
Base		100.0	

Demolition was the only trade present in all fifty-five projects. The range in the cost proportion of each trade indicates the variability in the size of the trades within the projects.

PROFILE OF ALL COST COMPONENTS IN A TENDER

The mean value of the cost proportion of the trades (see Table 6) was used to determine the cost profile of the twenty-one trades in the Priceable Building Work. This is shown in Figure 6. The finishing trade is seen to occupy the highest proportion of the Priceable Building Work at 21.2%.

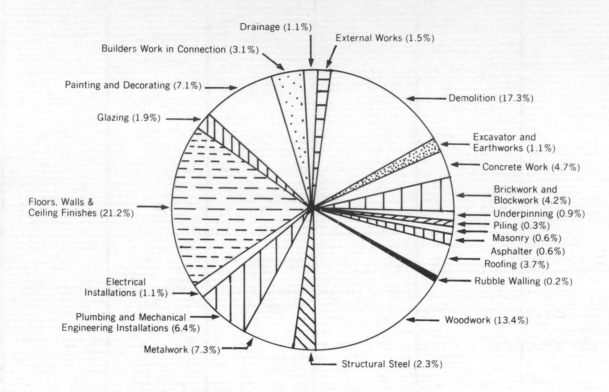

Figure 6   Cost profile of trades in a refurbishment project

5. VARIABILITY IN TENDER BID COMPONENTS OF A REFURBISHMENT TENDER

Several investigations were carried out with the object of determining the order of influence of each bid component on the overall success of a tender. The following inferences were drawn from the investigation:

BID COMPONENT WITH THE HIGHEST DEGREE OF INFLUENCE ON WINNING THE TENDER

The main component of Priceable Building work had the highest degree of influence on the lowest bidder winning the tender from the second bidder. In other words, the difference between the cost of the lowest bidder and second bidder was the greatest in this component of the tender. (This was expected by virtue of the size of this component in relation to the others). Comparing the lowest bidder and the third and fourth bidder, Preliminaries exerted a marginally higher influence than the Priceable Building Work on the lowest bidder winning the tender from his two competitors. Comparing the lowest bidder and the fifth and sixth bidder, Preliminaries stood out most clearly as a major influencing factor in the lowest bidder winning the tender. Never at any time did mark-up emerge as the most influential factor in the winning of a tender.

ORDER BETWEEN BID COMPONENT POSITION AND BID POSITION

It was found there was a significant relationship between the cost of the main components of mark-up, preliminaries, priceable building works and the order of the bids. A marginally less significant relationship was applicable to the other main components of Fixed Price Allowance and the Anticipated Return. There was no significant order in the position of the component of daywork and the position of the tender bid.

It can be concluded that when a bidder is in position 1, ie. if his bid is the lowest, it is quite likely that his bid component position is also lowest when compared to the bid components of his competitors.

INDEX OF VARIABILITY IN BID COMPONENTS

The following tables present the results of the statistical analysis to determine the variability in the Tender Bid Components of refurbishment works. They have been derived by comparing the tender build-ups supplied by the collaborating firms. Differences in estimating and tendering methods have been rationalised (as described in Section 3) so that the measures of variability indicate actual differences in costs submitted by competitors over a sample of 55 projects.

Index of variability in main components of a tender

Table 7 Index of variability in the cost of main components of the tender

Main component of tender	Index of variability
Priceable Building Work	100.0
Preliminaries	33.0
Mark-Up	8.0
Daywork	0.5
Fixed Price Allowance	0.2
Anticipated Return	9.5

It can be seen that the Priceable Building Work is the most variable main cost component in the tender.

Index of variability in sub-components cost of preliminaries

Table 8 Index of variability in the cost of sub-components of Preliminaries

Sub-component of Preliminaries	Index of variability
Staffing	100.0
General Labour & Special Attendances	67.4
Scaffolding	35.7
Other Components	25.3
Plant	16.7
Temporary Supplies & Services	8.7
Insurances	4.0
Temporary Accommodation	2.9

It can be seen that the staffing sub-component is the most variable cost component within Preliminaries.

Index of variability in trade sub-components of Priceable Building Work

Table 9 Index of variability in the trade sub-components of Priceable Building Work

Trade sub-components	Index of Variability
Demolition and alteration	100.0
Floors, walls & ceiling finishes	75.0
Woodwork	42.9
Metalwork	13.4
Painting and decorating	11.8
Plumbing & mech. eng. installation	9.6
Concrete work	7.0
General builder's work in connection	5.6
Brickwork and blockwork	5.6
Roofing	5.4
External works	4.0
Electrical installations	3.0
Structural steelwork	2.5
Underpinning	1.9
Glazing	1.3
Excavator	1.1
Masonry	1.0
Drainage	0.7
Rubble walling	0.5
Piling	0.3
Asphalt work	0.3

It can be seen that demolition is the most variable trade in the Priceable Building Work.

Index of variability in all components of a refurbishment tender

Table 10 Index of variability in the cost of all components in the tender

Component of tender	Index of variability
Demolition	100.0
Mark-up to tender	97.9
Floor, walls and ceiling finishes	75.0
Woodwork	42.8
Preliminaries – staffing	36.8
Preliminaries – g. labour & atten.	24.1
Preliminaries – scaffolding	16.4
Plumbing & mech. eng. inst.	15.6
Metalwork	12.0
Preliminaries – other items	11.4
Painting and decorating	10.3
Concrete work	8.7
Roofing	7.0
Preliminaries – plant	6.3
Electrical installations	6.3
Brickwork and blockwork	6.1
Dayworks	5.4
General builder's work in connection	5.3
External works	5.2
Preliminaries-temp supplies & services	3.5
Structural steelwork	3.2
Underpinning	2.7
Fixed price allowance	1.9
Glazing	1.6
Masonry	1.6
Excavation and earthwork	1.4
Preliminaries – insurance and bonds	1.4
Preliminaries – temporary accommodation	1.3
Rubble walling	0.9
Drainage	0.9
Piling and diaphragm wall	0.3
Asphalt work	0.3
Anticipated Return	101.0

It can be seen that the two most variable cost components in the tender were demolition and the mark-up applied to the tender.

Inferences from the study of variability in tender bid components

The index variability shown in Table 10 implies that contrators tendering for refurbishment work will be subjected to higher risk and uncertainty in the pricing of demolition by virtue of its highest index of variability in cost.

The variability in mark-up, on the other hand, could be due to one of several causes. The mark-up could have been used by competitors as a means of reflecting risk or scope in the tender. Thus, if a bidder views the risk in the tender as high, he will mark-up the tender at a corresponding rate. Conversely, if another bidder sees scope in the tender he will reduce his mark-up accordingly, resulting in the higher variability observed. However, the variability in the mark-up could also have been caused by the differing general overhead levels of competitors, as well as varying degrees of interest in the project. It was not possible to ascertain definitively which of the possible causes had the greatest effect on variability in the mark-up, without knowledge of the work load of the various competitors at the time of tendering.

Several sub-components of Preliminaries, namely staffing, general labour and scaffolding, also featured high in the hierarchy of variable cost components, reflecting uncertainty in managing the project on site.

However, if Anticipated Return is regarded as a component of the tender, it ranks as the most variable component in the tender. This implies that bidding strategy, as reflected in the Anticipated Return, featured most strongly as a cause for variation in the tender.

6. SUMMARY

REASONS FOR HIGHER VARIABILITY IN TENDERS

The high inherent variability in refurbishment tenders has its origins in a number of factors.

Unsatisfactory tender documents

The unsatisfactory state of tender documentation made a substantial contribution to the higher variability in the tender bids for refurbishment work. Amendments to the Standard Forms of Contract and the SMM and the ambiguous work descriptions within the tender bills, created different responses from competitors. A risk averse competitor resorted to incorporating a risk allowance in his tender, whilst a risk loving competitor made scope deductions from his tender in anticipation of claims from the contentious items. Other competitors sought to qualify their tender. The nett effect of these different responses is the high variability in the tenders.

High reliance on domestic sub-contractors' quotations

The profile of the lowest tender bid for refurbishment projects in Figure 4 indicated that about 64% of the Priceable Building Work in refurbishment tenders is based on domestic sub-contractors' quotations. The Priceable Building Work component of the tender also emerged as the most variable cost component in the tender.

A further investigation was conducted to determine whether competitors were using common domestic sub-contractors' quotations. If they were, then, the variability in Priceable Building work would have resulted from the builder's own work priced in-house by each competitor.

A total of 27 projects whereby information on a contractor's list of domestic sub-contractors from whom quotations had been invited, together with actual quotations used in the tender, were examined for the two trades showing the most variable cost difference between competitors.

For demolition, of the twenty-seven projects examined there were thirteen (48.1%) projects in which any two contractors invited quotations from one common sub-contractor. There were five (18.5%) projects in which any three contrators invited quotations from the same sub-contractor. However, there were only three (11.1%) of the twenty-seven projects in which any two contractors used the same sub-contractor's quotation in their respective tenders. The total pool of demolition domestic sub-contractors within these twenty-seven projects stood at forty-two. There were thirteen contractors who met irregularly in competition drawing from this pool. The chances that any two contractors would be inviting quotations from the same domestic sub-contractor are low. The chances of these two contractors using the same sub-contractor's quotation in their tenders are even lower. The same conclusion was drawn when alaysing the metalwork. The pool of sub-contractors in this case was even higher at forty-five.

Variations in the Priceable Building Work would, therefore, to a large extent, be caused by domestic sub-contractors quotations used in the respective tender.

Unequal sized competitors

The unique mix of unequal sized competitors, especially in smaller sized refurbishment projects, also had its effect on variability in the tender bids submitted. This is due to the varying 'competitive advantage' one competitor may have over the other by virtue of size, efficiency, ownership of supply resources, managerial skills and the like.

Hence the higher variability in the tender bids would be more pronounced in projects where the proportion of prime cost, provisional and contingency sums are high in the tender leaving the unequal sized competitors to compete mainly on Preliminaries and mark-up.

CONCLUSION

Consistent and excessive variability in tender bids leads to the conclusion that the basic mode adopted for the preparation of tenders, and the subsequent appointment of a contractor to execute the work, are fundamentally unsound.

Unless risk and liability can be reduced through better specification of the required work in the tender bills or shared between the client and contractor through the use of non-traditional approaches to contracting, such as fee or management contracts, the higher variability in tender bids for refurbishment work will remain a feature in the current estimating and competitive tendering climate.